Subtra

by Ann Corcorane

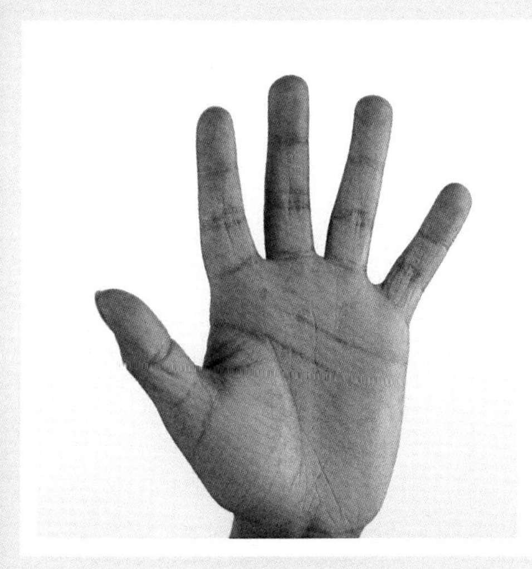

Consultant:
Adria F. Klein, Ph.D.
California State University, San Bernardino

capstone
classroom
Heinemann Raintree • Red Brick Learning
division of Capstone

We subtract when we take away.

Subtracting is fun!

There are 4 trucks.

Take away 1.
Now there are 3 trucks.

There are 5 pencils.

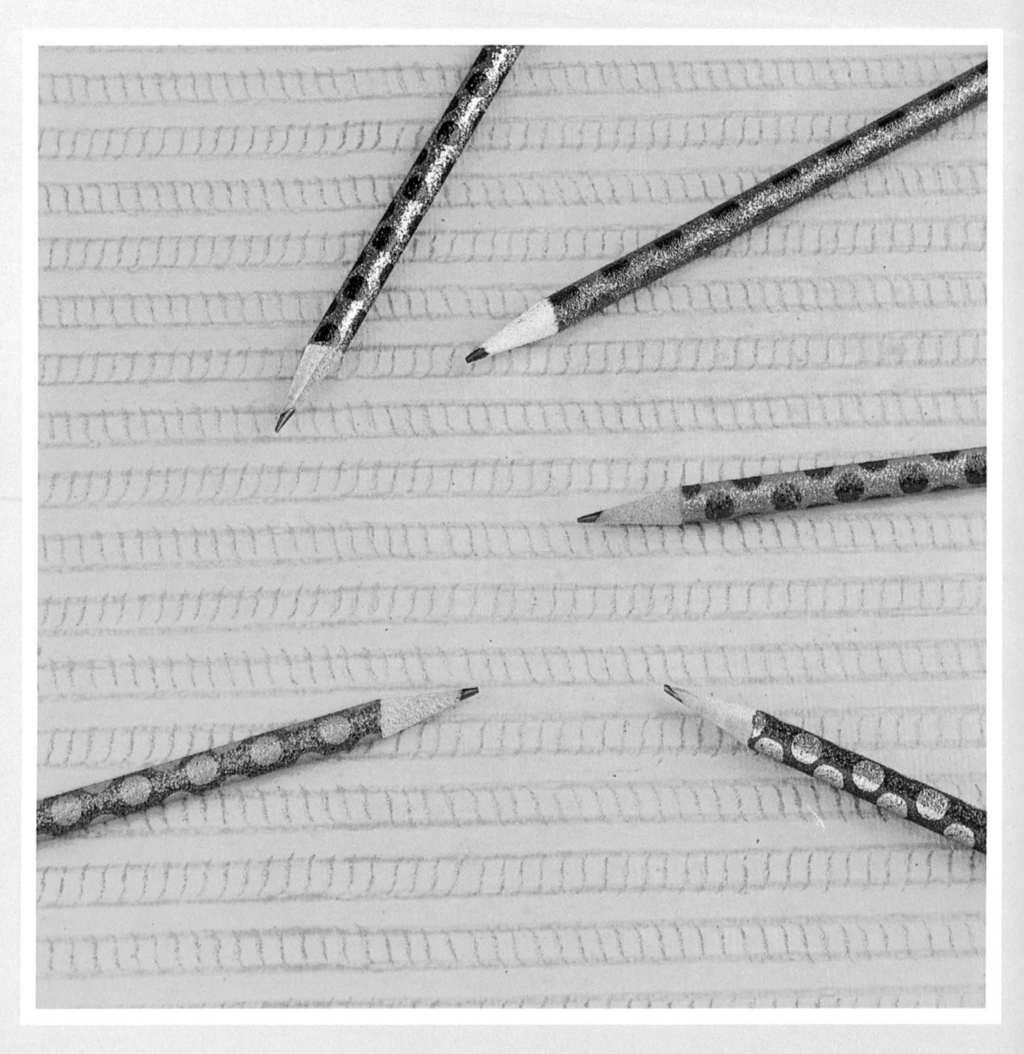

Take away 3.
Now there are 2 pencils.

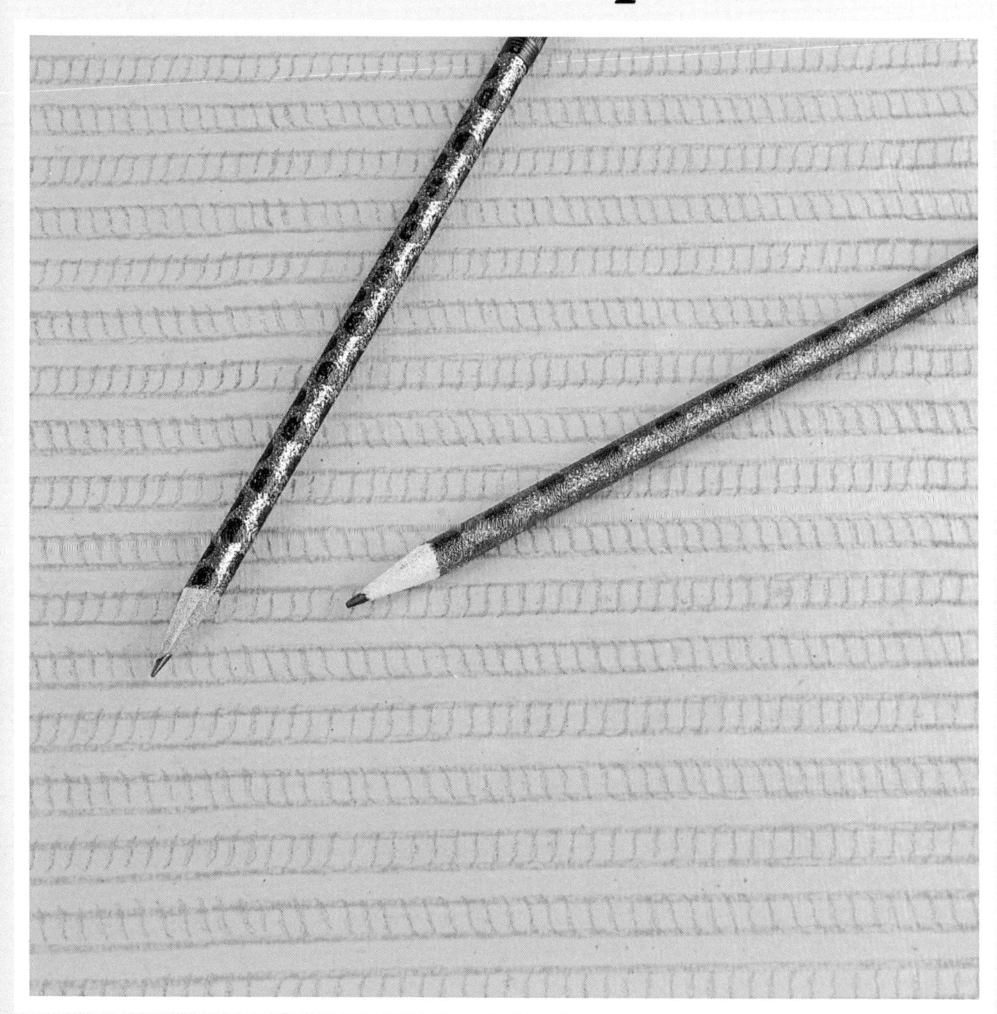

There are 2 frogs.

Take away 1.
Now there is 1 frog.

There are 8 shells.

Take away 4.
Now there are 4 shells.

There are 10 flowers.

Take away 5.
Now there are 5 flowers.

There are 6 teddy bears.

Take away 6.
There are 0 teddy bears.

There are 5 dolls.
Take away 2.
How many are left?

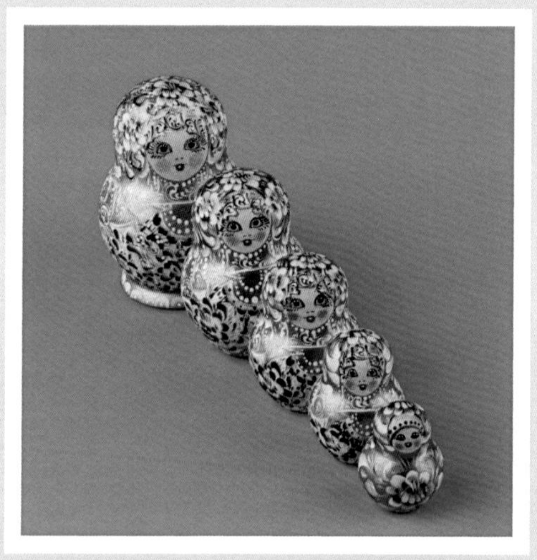